Ricardo Marques Couto

Interactions between bee and parasitoid: behaviour and manipulation

Ricardo Marques Couto

Interactions between [illegible] and processes, behaviour [illegible]
[illegible]

Ricardo Marques Couto

Interactions between bee and parasitoid: behaviour and manipulation

Research and teaching using the trap-nest technique

ScienciaScripts

Imprint

Cover image: www.ingimage.com

This book is a translation from the original published under ISBN 978-613-9-64393-6.

Publisher:
Sciencia Scripts
is a trademark of
Dodo Books Indian Ocean Ltd. and OmniScriptum S.R.L publishing group

120 High Road, East Finchley, London, N2 9ED, United Kingdom
Str. Armeneasca 28/1, office 1, Chisinau MD-2012, Republic of Moldova, Europe
Printed at: see last page
ISBN: 978-620-7-74192-2

Index

CHAPTER 1

The trap-nest technique and research with solitary bees.

1-Lone bees collecting oil.

Contrary to what many people think, most bees are not social, nor do they display aggressive behaviour. Approximately 85% of bee species are solitary, i.e. they live alone and there is no division of labour between the females of the same generation. These bees do not produce honey, royal jelly, propolis or wax. However, they play an essential ecological role in the maintenance of many plant species (Alves dos Santos 2004). In Brazil, of the 1678 known bee species, solitary bees are also the majority (Moure *et al.* 2007).

Although just as important as social species in their fundamental roles within ecosystems, solitary species have been proportionately less studied to date. This lack of knowledge is the result of at least two factors: firstly, because the populations of solitary species are often sparse (Danks 1971) and, secondly, because of the difficulty in locating their breeding (nesting) sites (Jayasingh & Freeman 1980).

In this work, we will investigate the solitary bee *Centris analis,* easily found on the "stalks" of the acerola fruit, from the Malpighiaceae family, which has oil-producing glands (elaiophore) on its flowers. The interesting thing is that these bees collect this oil. Why do they do this?

There is a long history of evolutionary interactions between Centridini and oil-producing plants, especially Malpighiaceae (Anderson 1979; Vogel 1990). Surveys in tropical biomes, especially the Cerrado, suggest that bees from the Centridini tribe show high local diversity (Carvalho 1990; Pedro 1994; Silveira & Campos 1995), where Malpighiaceae is also well represented

(Rizzini 1997).

Females of the tribe Centridini have specialised structures on their legs or sternum to collect, manipulate and transport oil. These structures are usually related to the different types of elaiophores (the oil-secreting glands), whether they are located in the epidermis (epithelial) or in specific trichomes (Vogel 1988; Buchmann 1987).

The oil collected is used by these bees as their own or larval food, along with pollen (Vinson *et al.* 1997; Camillo 2005), and also in the construction of brood cells (Hiller & Wittmann 1994; Camillo 2005). According to Neff & Simpson (1981), Buchmann (1987), Pereira et *al.* (1999), Aguiar & Garófalo (2004), floral oil is added to building material (sand, soil, sawdust, plant fragments) to make cells in various species of *Centris;* the four species of *Tetrapedia* studied by Camillo (2005) also associated the oil collected with sand to build their nests.

The transparent and viscous appearance of the substance observed in the brood cells of *C. trigonoides* (Simpson *et al.* 1977) suggests that floral oil may also be used by this species to coat the cell internally, as has been observed to occur at least in *C. analis (*Jesus & Garófalo 2000), *C. vittata* (Pereira *et al.* 1999) and *C. tarsata* (Aguiar & Garófalo 2004).

Several authors have suggested that coating the nest closure with oil provides protection against parasites that attack completed nests (Pereira *et al.* 1999; Jesus & Garófalo 2000; Gazola & Garófalo 2003) or against nest usurpation by co-specific females (Vinson & Frankie 2000). Gazola & Garófalo (2003) observed that, on several occasions, *Leucospis cayennensis* Westwood was unsuccessful in breaking through *THE* closing wall of the *C. analis nest,* probably due to the hardening provided by the oily material deposited there.

Simpson et *al.* (1977) and Seigler et *al.* (1978) identified the presence of p-acetate in the oily substance obtained from species of the genus *Krameria* (Krameriaceae). Vogel (1974, 1976a,b) identified the presence of acetic acid and palmitic acid in species of *Calceolaria* (Calceolariaceae). In Costa Rica, Vinson *et al.* (1996) identified 16 species of *Centris* visiting *Byrosonima crassifolia* (Linnaeus) (Malpighiaceae) whose oil showed a mixture of mono-, di- (80-85%) and tri-glycerides (10%) and fatty acids (5%).

Cane *et al.* (1983) identified in *Lysimachia ciliata* (Linnaeus) (Primulaceae) components analogous to the secretions of the Dufour's gland of *Macropis nuda* (Provancher) (Hymenoptera: Melittidae); among these components were observed palmitic acid, aldehydes with carbon chains of 14 and 16 carbons (C14 and C16), acetates (C18-C22), alkanes (C23, C25, C27 and C29), alkenes (C25 and C27), as well as various esters.

C- Biology of *Centris analis.*

The nesting behaviour of *C. analis* in trap nests was described by Jesus (1995) in a study carried out on the campus of the University of São Paulo in Ribeirão Preto, SP. According to the author, the bees used plant material and an oily substance of unknown origin to build their nests. The plant material was used to build the bottom wall of the cells, cell partitions and the closing wall of the nests. The oily substance was used to coat the inside of the cell bottom, the cell walls and the outside of the nest closing wall. The cells were filled with pollen and nectar and those where females were produced received significantly more pollen deposits than those where males were produced. With regard to nectar, there was no significant difference between the number of deposits in female and male cells.

Pereira *et al.* (1999), Jesus & Garófalo (2000), Gazola & Garófalo (2003) have suggested that coating the closure of *Centris* nests with an oily substance provides greater protection against parasites and predators that attack completed nests. Gazola & Garófalo (2003) observed that, on several occasions, the parasite *L. cayennensis* was unsuccessful in breaking through the closing wall of the C. *analis nest,* probably due to the oily substance deposited there. According to Evans (1966) and Medler (1958), the presence of the vestibular cell (a structure usually present in *C. analis* nests that precedes the closing wall) also reduces the success of *L. cayennensis* attacks.

For the trap nests used, the number of brood cells in completed nests varied from 1 to 4, but the presence of 3 cells was observed more frequently. The cells were built in a linear series, usually followed by a vestibular cell, which corresponds to the space between the closing of the last brood cell and the closing of the nest (Figure 1).

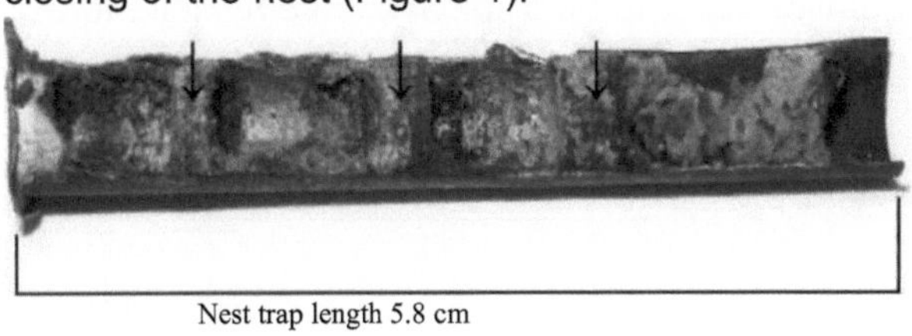

Figura 1- Longitudinal section of trap nest showing brood cells (A,B,C,D).

The cells used to produce females ranged in length from 0.88 to 1.21 cm and were significantly larger than those used to produce males, which ranged in length from 0.83 to 1.18 cm. The length of the buccal cell was related to the number of brood cells built in the nest and varied from 0.37 cm in a nest with 4 brood cells to 3.52 cm in a nest with only 1 brood cell. The time taken to complete a cell varied from 1 to 5 days, but most were completed within 1 day.

Females were produced in the innermost cells and males in those closest to the nest entrance. The longest period of activity for a female was 30 days, and the maximum number of nests built by a female was 3 and the maximum number of cells was 9. *Leucospis cayennensis* (Hymenoptera, Leucospidae), *Coelioxys* sp. (Hymenoptera, Megachilidae) and *Mesocheira bicolor* (Hymenoptera, Apidae, Ericrocini) were the parasitoids that attacked the nests of *C. analis,* with the former being the most frequent and the latter the rarest.

3-The trap-nest technique and the study of solitary bees.

The study with the solitary bee *Centris analis* was carried out on the campus of the University of São Paulo, located in the municipality of Ribeirão Preto, SP, between the coordinates 21° 05'- 21° 15' S and 47° 50' - 47° 55' W. The relief has altitudes ranging from 530 to 675 m, in an area of approximately 573 ha. The soil is characterised as eutrophic purple latosol, with a clayey or very clayey texture. The vegetation used to be tropical sub-caducifolia forest, but the region was intensively exploited, initially for coffee growing and more recently for sugar cane, which together with the colonião grass forms the current cover.

Following the methodology presented by Serrano & Garófalo (1978), we used trap nests made from black cardboard, in the form of small tubes, with one end closed with the same material, with a diameter of 0.6 cm and a length of 5.8 cm. These trap nests (NA) were inserted into holes made in wooden boards (n=08) (55 NA each) 5.0 cm wide, 30.0 cm long and 12.0 cm high (Figure 2).

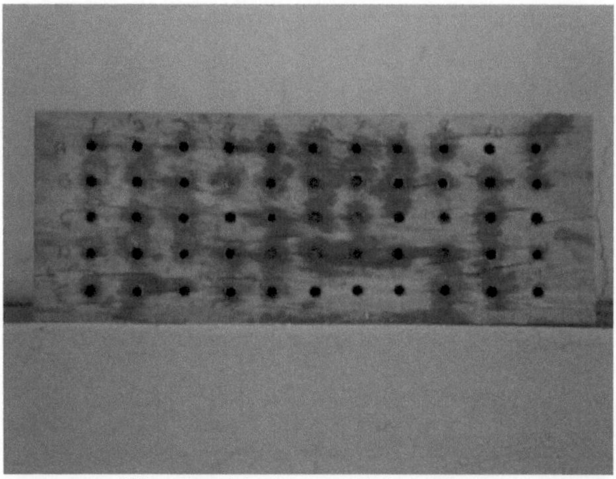

Figura 2- Wooden block containing 55 trap nests (NA).

The plates were placed on shelves built next to the Ecology laboratory of the Biology Department of the Ribeirão Preto School of Philosophy, Sciences and Letters (Figures 3). During the studies, *C. analis* females were marked with ink on the thorax and the nesting activities of each one were recorded daily.

When activities were restricted to inside the nests, they were recorded using an otoscope.

Figura 3- - Nesting site on the campus of the University of São Paulo in Ribeirão Preto.

The completed trap nests were removed from the plates and placed in test tubes, which were closed and kept in the laboratory at room temperature. Once the individuals had emerged, the nests were opened and their contents analysed to determine the sex ratio and dead immatures. Parasitoid attacks were monitored using direct observations. After identification, the adults of *C. analis* were released into the field.

4- Parasites or Parasitoids?

Parasites are organisms that live in association with the host from which they derive the means for their survival, usually by weakening the host organism, a process known as parasitism. Parasites have a significant impact on host activity and habitat preference (Poulin 1994, Klein 2003). According to

Thomas *et al.* (2002), the manipulation of the terrestrial grasshopper *Meconema thalassinum (*DeGeer) (Tettigoniidae) by the nematode worm *Spinochordodes tellinii* (Camerano) (Chordodidae) causes a drastic behavioural change in the parasitised insect, causing the grasshopper to throw itself into the water so that the parasite can emerge and reproduce.

Parasitoids, on the other hand, are organisms that need a host to complete their development, which is killed during the process. These parasitoids are generally used as biological control agents (Michener 2000). Parasitoids also cause behavioural changes in their hosts (Norton & Moore 1993), as observed in *Platyprepia virginalis* Bvd. (Lepidoptera: Arctiidae) when parasitised by *Thelaira americana* Brooks flies (Diptera: Tachinidae), which alter their distribution and preference for the host plant.

Parasitoids have developed various mechanisms to keep their hosts in suitable conditions, not only during larval development, but also during pupation. The larva of *Cotesia glomerata (*Linnaeus) (Braconidae) consumes no more than 50-60% of the food assimilated by the host (Slansky 1978).

Parasitoids depend on the host not only to fulfil their nutritional and metabolic needs, but also to reduce the risk of premature death through predation. The survival of immature parasitoids depends on the "performance" of the host, in particular the ability to avoid natural enemies. Any alteration to the parasitised host that increases the risk of predation should be avoided until the parasitoid has completed its development (Thomas et *al.* 2005).

Solitary and pre-social Hymenoptera (Apoidea:Vespoidea) are often hosts for many families of flies (Diptera), such as Anthomyliidae, Conopidae, Phoridae, and Sarcophagidae (Bohart & Menke 1976, Coville & Griswold 1983, 1984, Wcislo 1990, Paxton et *al.* 1996, Andrietti et *al.* 1997, Polidori et *al.* 2001). Flies from the Conopidae family are probably the most important parasitoid dipterans of bees, which are attacked when they are foraging on flowers.

There are many records of associations between *Physocephala* and various genera of bees, such as *Anthidium, Anthophora, Eucera, Megachile, Halictus, Apis, Xylocopa, Bombus and Eulaema* (Linsley 1958, Hurd 1978, Katayama & Maeta 1998, Schmid-Hempel 2001, Otterstatter *et al.* 2002, Rasmussen & Cameron 2004).

The first observations of the association between *Centris analis (Fabricius)* and *Physocephala* species were made on the campus of the University of São Paulo, Ribeirão Preto, SP. From a sample of 26 females and 9 males of the host species found dead inside empty trap-nests, 35 parasitoids belonging to nine *Physocephala* species emerged (Santos *et al.* 2008). The parasitoids belonged to eight species: *Physocephala soror* Krõber (n= 2), Physocephala *ipunctata* Macquart (n = 1), Physocephala *inhabilis* Walker (n = 5), Physocephala *rufithorax* Krõber (n = 5), Physocephala *cayennensis* Macquart (n = 2), Physocephala *aurifrons* Walker (n = 1), Physocephala *bennetti* Camras (n = 1), Physocephala *spheniformis* Camras (n = 3).

Figura 4- Female *Centris* bee *analysed* as marking on thorax found dead inside trap nest.

5- Atypical behaviour of oily substance deposits.

In this study, 11 C. *analis* females parasitised by the conopid had their oil deposition behaviour quantified, analysed and discussed. All of them were found dead at the nesting site, inside trap nests, suggesting that these parasitised females moved to a safe place before dying, according to Muller

(1994). Due to the life cycle of conopids, the larva has to pupate and hibernate inside the host to complete its development, and a safe place to withstand extremes of temperature, dehydration, humidity, predators and hyperparasitism is crucial for the survival of the parasitoid pupa (Brodeur & Boivin 2004).

The 11 parasitised C. *analis* females built 37 nests with brood cells and 42 nests that were only closed (empty), without brood cells. All the nests with brood cells received an extra deposition of oil at the close of the last cell partition (Figure 5), in a similar way to the deposition of oil at the close of the nests.

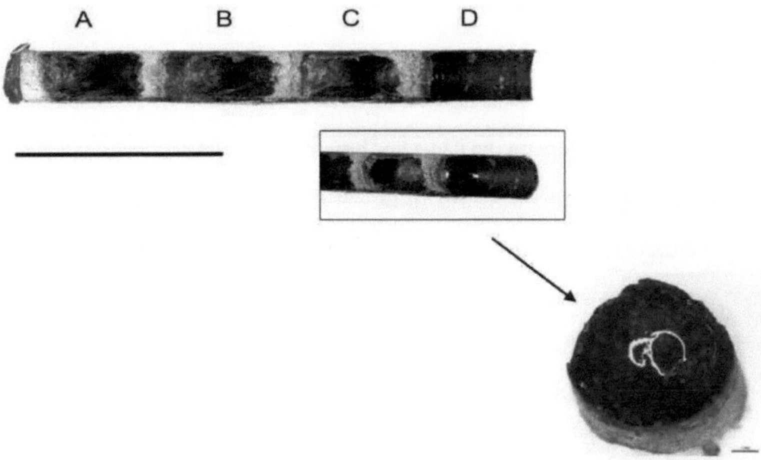

Figura 5- Extra deposition of oily substance on the closure of the last brood cell.

With regard to the length of time the parasitised females remained at the nesting site after closing their nests, these females spent an average of 5.4 days depositing extra oil (n = 134) in previously closed nests, of which 26 extra deposits (19.4%) occurred when the females deposited oil on their own nests, i.e. on 15 of them, equivalent to 40.5%. In the other 108 extra depositions (80.6%), the oil was placed in the closure of nests built by other females, of which 53 nests (68.8%) received just one extra deposition, 18 (23.4%) received two depositions, five (6.5%) received three depositions and only one nest (1.3%) received four extra depositions.

Table 1. Number of nests that received one, two, three and four extra deposits of oil in their closures.

	No. of nests that received 01 extra deposit	**No. of nests that received 02 extra deposits**	**No. of nests that received 03 extra deposits**	**No. of nests that received 04 extra deposits**
Oil deposits in their own nests	05a	09a	01b	-
Oil deposits in nests built by other females	53a	18b	05c	01

- Values followed by different letters on the same line differ statistically.

It could be suggested that the nests of *C. analis that* receive an oily substance when closing and at the same time in the last partition may be more protected from attacks by the parasitoid *L. cayennensis* because the cell would be doubly protected by the surfaces that have received this substance.

According to Gazola & Garófalo (2003), this parasitoid has two attack tactics when entering the nest: a) the parasitoid introduces its ovipositor through the nest's closing plug until it reaches the last cell partition. In these cases, it is easier to reach the inside of the last cell, as only the closure would have an oily substance; b) the parasite breaks through the closure plug, enters the nest and walks until it reaches the wall of the last cell partition. If it is coated with oil, the parasitoid will have a harder time.

On several occasions, Gazola & Garófalo (2003) observed that L. *cayennensis was* unsuccessful in breaking through the closing wall of the C. *analis nest,* probably due to the hardening provided by the oily material deposited there. In studies carried out in Costa Rica with *Centris* species, Vinson et *al.* (2006) suggested that the oily substance collected by the females was not only used for provisioning the nest, but also for protecting it.

In the same way that the oily substance protects the nest from parasite attacks (Gazola & Garófalo 2003), empty cells also help to reduce the rate of parasitism (Sakagami & Laroca 1971, Wcislo & Orozco 1993). In this study, the 11 females of C. parasitised and analysed closed 42 empty trap-nests without brood cells.

5.1- Leucospis attacks.

Before discussing the attacks of *Leucospis cayennensis* on *C. analis* nests, it is important to know the strategies that ensure its survival. The occurrence of *L. cayennensis* parasitising *C. analis nests* was first reported in Brazil by Roubik (1989), Jesus & Garófalo (2000) and Gazola & Garófalo (2003). Some attacks by *L. cayennensis* were carried out without the parasite breaking through the closure, while in others it occurred after the wall had been breached.

In those attacks that occurred in the anterior region of the nests, the females had two procedures: a) in nests with 4 brood cells, with a reduced vestibular cell, the parasite introduced its ovipositor through the nest's closing plug, passing through the last cell partition, reaching the inside of the brood cell; b) in nests with 2 and 3 brood cells, the parasite broke through the closing plug, entered the nest, and walked until it reached the wall of the last cell partition when it then introduced its ovipositor inside the cell.

These results are certainly related to the length of the vestibular cell, which is a structure generally present in *C. analis* nests with less than 4 brood cells (Jesus 1995). As the length of the ovipositor of *L. cayennensis* females is

5.09 ±1.16 mm (Menezes 1992) and in *C. analis nests* containing 4 brood cells the vestibular cell measures 3.70 mm (Jesus 1995), the parasite, in these cases, has no difficulty in reaching the last cell in the nest with its ovipositor and parasitising it. On the other hand, in nests containing up to three brood cells, the vestibular cell is longer than the parasite's ovipositor. So, in these cases, the parasite cannot reach the last brood cell in the nest and, consequently, the attack is unsuccessful.

Although it has been proposed that the presence of the vestibular cell is related to defence against the action of natural enemies (Krombein 1967, Coville & Coville 1980, Coville 1982, Coville & Griswold 1984), there is no data to confirm this function. Results obtained by Gazola & Garófalo (2003) with *C. analis* suggest that the presence of the vestibular cell and the closing wall coated with an oily substance in the nests of *C. analis* reduce the success of attacks by *L. cayennensis.* However, according to Krombein (1967), *L. cayennensis* can also attack the host nest after breaking through the nest closure and, by entering the vestibular cell, reach the last cell with brood, managing to parasitise it.

In this study, the success of *L. cayennensis* attacks was higher in nests with 4 brood cells, since parasites emerged in all the nests attacked (n=12). In the case of nests with 3 brood cells, attack success was 46.1%, i.e. of the 13 nests attacked, in 6 of them the parasite managed to introduce its ovipositor through the cell wall and parasitise the nest (Table 2). In the other seven nests attacked, the presence of an oily substance on the closure of the last cell was noted, making it difficult for the ovipositor to pass through the closure wall and, consequently, for parasitism to take place.

It is worth noting that in the nests attacked by *L. cayennensis,* with 3 brood cells and which received an oily substance on the closure of the last brood cell, 6 adults emerged and one dead *C. analis* larva was found. This is important because it clearly shows that the attacks by L. *cayennensis were* unsuccessful.

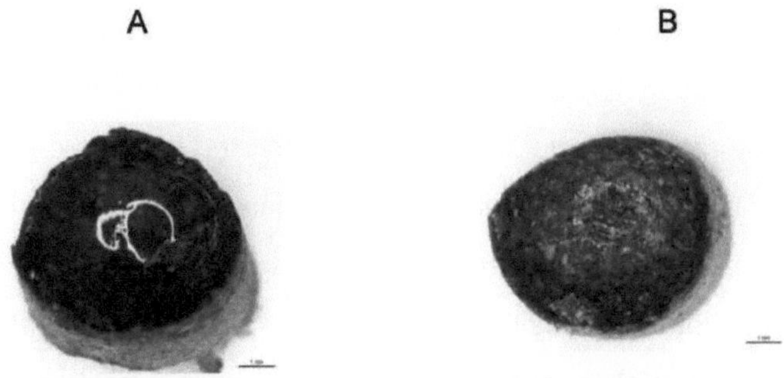

Figure 6- Nests with 3 brood cells: (A) Extra oil deposition in the last brood cell; (B) No extra oil deposition in the last brood cell.

Table 2 - Percentage of success of *Leucospis cayennensis* in nests with 1, 2, 3, 4 and 5 brood cells built at the nesting site

No. of cells/ Nest	**No. of nests built**	**No. of parasitised nests**	**No. of attack successes**	**(%) of success**
1	06	00	-	-
2	19	01	0	0a
3	39	13	6	46,1b
4	28	12	12	100c
5	02	01	1	100c
Total	94	27	20	74,0

- The results (%) of successful attacks followed by different letters in the same column differ statistically.

5.2- Emergencies, mortality and sex ratio.

Of the 37 nests and a total of 114 brood cells built by the parasitised females, there were 43 emergences (37.7%) and mortality (62.3%) in the egg, larva, pupa and parasite attack stages.

Of the emergences, the ratio of females to males stands out. There were 26 females and 17 males, giving a sex ratio of 1.5:1. In most panmictic populations, equal investment in both sexes is expected and the sex ratio is close to 1:1. The data from this study show that more females than males emerged from the offspring of females parasitised by the conopid and which died inside the NA. These females presented an oviposition rate of 1.35 eggs/day, close to the rate presented by the non-parasitised females obtained by Jesus (1995). These results show that the parasitoids (Conopidae:Physocephala) did not interfere with the laying of the host.

On the other hand, parasitism, according to Carver *et al.* (1999), causes morphological, physiological and behavioural changes in hosts and can influence egg production. For example, in the beetle *Tenebrio molitor* (Linnaeus) (Tenebrionidae) infected by the worm *Hymenolepis diminuta* (Rudolphi), there are qualitative and quantitative changes in oocyte development (fecundity) and egg laying (fertility) (Hurd & Arme 1986). Egg production is delayed due to a reduction in vitellogenin synthesis (Webb & Hurd 1996, Hurd & Arme 1986) and protein deficiency in the ovary (Hurd & Arme 1986). Although the "performance" of the ovary is regulated by factors originating in the host (Major *et al.* 1997), the fall in vitellogenin production in the fat body is influenced by factors of parasitic origin (Webb & Hurd 1996).

5.3- Intraspecific parasitism (usurpation).

Another atypical behaviour shown by *C. analis* females was ovipositing in co-species nests. The behaviour of ovipositing in nests of the same species is called intraspecific parasitism (usurpation) (Field 1989). In his review of the occurrence of intraspecific parasitism, Field (1992) defined 6 types of parasitism, which can be found among bee and wasp species nesting solitarily or communally:

1-Brood parasitism, when the parasite replaces the host's egg with its own,

2-Theft of provisions from the host's nest,

3-Taking provisions from the host's nest, but then abandoning them,

4-Theft of provisions from the host before they are taken into the nest,

5-Possession of the host's nest structures,

6- Possession of the host's nest structures, after removing the host's cells.

The parasitised females (n=11) usurped 7 nests (intraspecific parasitism). Of the 7 usurped nests, in 5 of them the females broke through the closure, entered the nests, removed all the material inside (pollen and larvae) and nested in them. In the other 2 nests there was a physical dispute between the females and the usurpation occurred when the female in the nest was displaced by the other female.

The intraspecific parasitism observed in this study by the females parasitised by the Conopid is type 3. On the other hand, the parasitism observed by Jesus (1995), in a study carried out at the USP Campus in Ribeirão Preto, in which neither parasitised females nor atypical behaviours were identified, falls only within the first type defined by Field (1992).

The data reported here suggest that the behaviour of removing provisions from the host's nest and then abandoning them is an atypical behaviour displayed by *C. analis* females. Low food availability and low availability of nesting sites, two of the possible conditions that can favour the appearance of parasitic behaviour (Field 1992), were not observed during the study period and can therefore be ruled out.

6- Manipulated behaviour in hosts.

In bees, various mechanisms have been developed to protect their offspring from parasites, including aggressive behaviour and some features of nest architecture (Krombein 1967, Sakagami & Laroca 1971, Tepedino *et al.* 1979, Wcislo & Orozco 1993). Aggressive behaviour was reported by Kapil & Dhaliwal (1968), when they observed that females of *Xylocopa fenestrata*

(Fabricius) (Apidae) provide protection for their offspring by applying a repellent substance to the entrance of their own nest when harassed by external agents.

However, when considering parasitoids, the behaviours identified in hosts are different. Hosts can show behavioural, physical and physiological changes that are used to escape from their predators (Edmunds 1974, Curio 1976, Endler 1986, De Jong-Brink et *al.* 1997, Combes 2001, Moore 2002, Thomas et *al.* 2005). Some parasitoids control the behaviour of their hosts to avoid attacks by predators, and changes in the distribution and activities of hosts are the most common behaviours manipulated by parasitoids (Norton & Moore 1993).

According to Brodeur (1994), hosts often present adaptations that avoid natural enemies, which can be exploited by immature parasitoids to promote their own survival. Changes in host behaviour are more likely to benefit the parasitoid because the parasitised host has no reproductive future.

Norton & Moore (1993) identified two main categories of changes in host behaviour: changes in habitat preference and changes in activity. Parasitised hosts can change their habitats, being found at higher altitudes (Stamp 1981, Brodeur & McNeil 1991, Brodeur & McNeil 1992) or in places that are more difficult to find (Brodeur & McNeil 1989, Muller 1994, 1996, Reitz & Nettles 1994). These differences are usually associated with thermal, geostatic and photostatic conditions (Brodeur & McNeil 1990, Horton & Moore 1993).

Brodeur & McNeil (1989, 1992) observed that the potato aphid *Macrosiphum euphorbiae (*Thomas) (Hemiptera:Aphididae), when parasitised by *Aphidius nigripes (*Ashmead) (Hymenoptera:Aphidiidae), alters the location of the aphid colony, moving it from the underside of the leaf to the upper side, where hyperparasitism rates are lower. According to Reitz & Nettles (1994), the parasitoid *Eucelatoria bryani* Sabrosky (Tachinidae:Diptera) regulates the behaviour of the larva of its host *Helicoverpa zea (*Boddie) (Lepidoptera:

Noctuidae), which normally penetrates the soil to pupate.

Parasitised hosts burrow into the soil earlier than the non-parasitised host, causing the larvae of the *H. zea* host to burrow into the soil more quickly, thus ensuring pupation in an environment with greater protection from attacks by natural enemies and lethal temperatures.

In *Bombus terrestris* (Linnaeus) (Apidae) this parasitic association has been studied and has shown an increase in the mortality of worker bees (Schmid-Hempel & Schmid-Hempel 1990), changes in foraging behaviour (Heinrich & Heinrich 1983, Muller & Schmid-Hempel 1993) and a reduction in colony size (Muller & Schmid-Hempel 1992, MacFarlane *et al.* 1995).

The behavioural changes observed were: a reduction in foraging activity, a change in the pattern of pollen collection and a change in the preference of flowers for foraging (Schmid-Hempel & Schmid-Hempel 1990). When these bees were parasitised, they had a preference for colder temperatures (Muller & Schmid-Hempel 1993), and also showed a tendency to leave the colony (Schmid-Hempel & Muller 1991, Muller & Schmid-Hempel 1993). According to Schmid-Hempel (1991), when bees of the genus *Bombus are* close to death, parasitised individuals bury themselves in the ground more often than non-parasitised individuals, reducing the chances and probability of their bodies being found by predators. This behaviour benefits the parasitoid and has no adaptive value for the host.

Muller (1994) suggests that parasitised bees should move to safe places before they die. In the present study, females of *C. analis* (n=11), parasitised by the conopid, were found dead at the nesting site (inside trap nests placed on perforated plates under cover, sheltered from the sun and rain), suggesting that these parasitised females had moved to a safe place before dying.

The data reported here show that C. *analis* females parasitised by the conopid showed behaviours involving the deposition of an oily substance that determined the success of *L. cayennensis* attacks. Several authors have suggested that placing oil on the nest closure provides protection against parasites attacking completed nests (Pereira *et al.* 1999; Jesus & Garófalo 2000, Gazola & Garófalo 2003) or against nest usurpation by co-specific females (Vinson & Frankie 2000).

The depositions of oily substance by the females parasitised by the conopid increased the difficulties for the success of the parasite *(L. cayennensis).* It's worth pointing out that by defending their nests against these parasites, they (the females) would be ensuring a new generation of the conopid. This assumption could be even more abusive if we take into account that we could infer the occurrence of a competition between the parasites, i.e.: the conopid would be (through the host female) trying to secure more substrates for itself, while the conopid would be trying to secure more substrates for itself.

Leucospis tries to reduce the substrate to be used by the Conopid (Couto & Camilo 2014).

CHAPTER 2

Researcher who teaches and teacher who researches.

1- Using the nest-trap technique with primary school students.

Using the trap-nest technique enabled students to learn about bees through their life cycle, thus developing other skills - such as: observing, identifying, comparing, analysing, formulating hypotheses, synthesising, judging, among others. To do this, it was taken into account that learning these skills requires differentiated activities (Viella 2006) and, in this case, the educational objectives of Bloom *et al.* (1973) should be considered when planning teaching practices, seeking to reach more abstract and complex levels (Vianna 1982).

In this context, by subjecting students to differentiated planning and organisation, anticipating errors and systematising execution, it is the activity that leads students to improve their academic-scientific performance, which is reflected in their individual development (Couto & Castro 2014).

The students' knowledge that all bees are social and aggressive was the starting point for drawing up a research plan to overcome these preconceptions. In eight weekly meetings (24 hours of guidance) the students used the technique of building trap nests, identifying and attributing meaning to the characteristics of the life cycle of the *Centrisanalis* solitary bee. The opening of the trap-nests required a great deal of careful theoretical and technical preparation, preceded by and integrated into a project that guides it, known as "Little Scientist", from the House of Science programme at the Hemocentro/USP in Ribeirão Preto. Reflecting on the results, it was possible to problematise important concepts such as: feeding behaviour, parental care, adaptations, holometabolous development.

1.1- Parental care and holometabolic development.

As well as building the nest, the female is also responsible for provisioning. The opening of the trap nests showed a relationship between the deposition of pollen (food) and the development of the larva, suggesting that pollen was the main source of larval food. The trap-nest technique made it possible to identify the bees' specialised feeding behaviour, as well as to better understand holometabolous development. It is also important to mention the advantages of using insects as models for teaching practice: they have short life cycles and are easily manipulated.

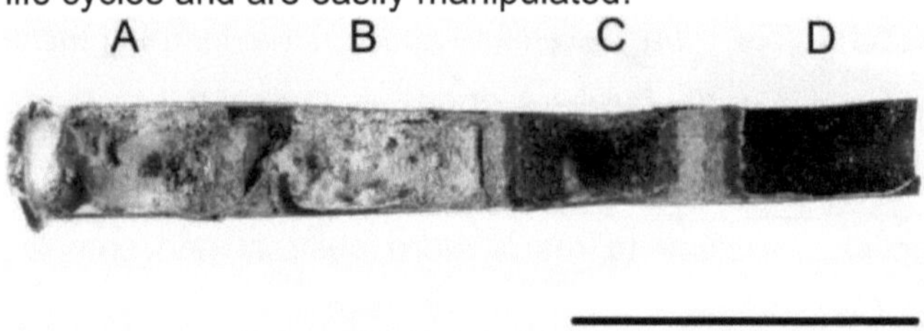

Figura 7- Longitudinal section of trap nest containing pollen in brood cells A and B.

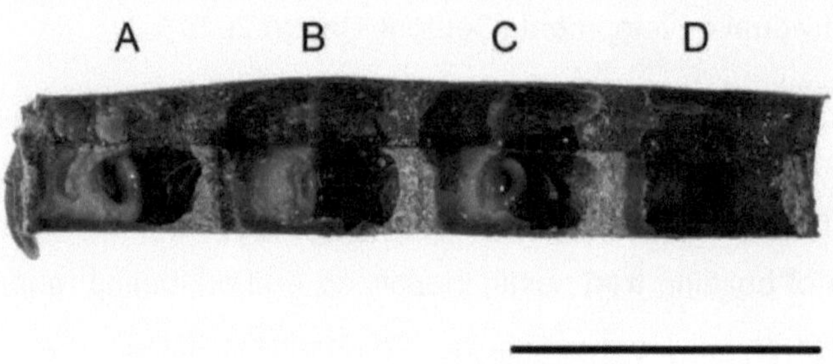

Figure 8-Longitudinal section of trap nest containing 3 larvae in brood cells A, B and C.

1. 2-Life cycle and adaptations.

During the meetings, the students' expressions showed that they had assimilated scientific language and articulated concepts, especially holometabolous development, which led to new questions and reflections on the life cycle and adaptations of bees.

- Student 1: As bees take a long time to develop because they go through the egg, larva and pupa stages, it's very likely that when they emerge from their nests as adults, the female parent has already died. Why is that? Why? Did she go to all that effort to die?

- Student 2: These holometabolous insects spend a good part of their lives in the nest, developing. Flying only takes up part of their cycle!

In order to work on these issues, it was necessary to construct the concept of adaptation as a historical approach, associated with the life cycle. In other words, we investigated the feeding behaviour of insects based on the evolutionary scale, starting with the most basal species, such as dragonflies (hemimetabolous), which do not select food for their immatures, i.e. the oviposition behaviour of the females is not related to the feeding process of the immatures. When they hatch, the immatures need to forage and eat whatever food is available. As a result, they are omnivores. Therefore, in these more primitive species, the females do not select sites to oviposit (lay their eggs) and, as a rule, immatures and adults feed opportunistically.

Then we investigated insects that undergo complete metamorphosis (Holometabolous) and discovered parental care. In the process of evolution they are known as beetles, butterflies, flies, bees, wasps and ants. In these insects, as the responsibility of the adult females for choosing the diet for the immature (offspring) increased, there was the possibility of the emergence of food specialisation, since the females are responsible for selecting the nesting site (building the nests) and the food for the immature, since they move little or not at all.

Considering this historical context, it is clear that morphological specialisations and behaviour are present in the evolutionary process of holometabolous bees and have been fundamental to the success of this group.

Bees have only perpetuated themselves over time because they have chosen very well the locations for reproduction and the diet of their offspring. In other words, they ensured a safe development, without the danger of predators and lack of food, but they paid a high price: they died before meeting their offspring!

This discussion on evolution and holometabolous development could not be left out of the orientation. Adaptation occupies a central role as a phenomenon to be explained by the theory of evolution by natural selection, and the tendency of this concept to be considered polysemic, both in the specific domain of biology and in other domains of culture and language.

In research carried out with secondary and higher education students, according to Sepulveda *et al.* (2013), it is clear in the ontogenesis of the concept of adaptation as a finalist phenomenon, according to which adaptation is sufficiently explained by the end or goal it fulfils. In general culture and even in biological discourse, we also find adaptation understood as a state of being, used in a-historical terms to designate the supposed propensity of organisms to have reproductive success. However, this use of the term has been criticised by evolutionary biologists and philosophers such as Mayr (2005), Sober (1993) and Burian (2005) on the grounds that the word adaptation should be used in historical terms to designate characteristics resulting from a process of natural selection.

It is on this stage of simulations and investigations that stimulating environments can form for the creation of models by the subject, in which each level is more complex and more specific than the previous one (Bloom et *al.* 1973).

In these situations, the subject is faced with a representation of reality, forcing them to formulate their own that fits in with the simulation, since science and biology teaching is still done in a descriptive way with a significant amount of vocabulary, making the learning process decontextualised and unmotivating (Krasilchik, 2005). According to Barbieri (1988), improving teaching involves clarifying the concept of experimental teaching. It also involves looking for methodologies that make this teaching feasible, taking into account that there is a close link between the conceptual basis and the path of content exposure. Defining what is meant by experimental teaching in biology, looking for suitable methods and a conceptual basis continues to come up against a major difficulty: the issue of teacher education and training for teaching biology in the context of the current characteristics of the school.When proposing a new way of teaching, the conceptions of learning and knowledge that underlie the model must be made explicit. It's not a question of destroying the students' previous conceptions, but of developing a teaching process that promotes the evolution of their ideas. In other words, rupture does not necessarily mean discarding (Ausubel, 1982). Thus, teaching cannot be conceived as a simply linear process in which new concepts are introduced sequentially, but rather as a process in which the teacher must also plan and develop frequent situations in which concepts that have already been covered are revisited and reworked in new ways, establishing new conceptual relationships in order to provide students with the conditions to apply, expand and consolidate those ideas, i.e. scientifically accepted ("correct") ideas.

2- An innovative proposal: Trap-nests, lesson planning and assessment.

According to Barbieri (2001), in teaching practice, the lesson plan is an instrument for the teacher's decision and should include his or her learning hypotheses gathered in an investigation plan, as a proposal to be tested. It is similar to a research project in that this plan represents a section that explains the learning hypotheses to be tested in the lesson. In keeping with the qualitative method, in the contaminated environment of the classroom, in the sense of the interactions established between the subjects, the plans are validated in the evaluation process (Minayo 1992).

The plans - organised to project expected actions and find unexpected ones - reveal a complex network of links between simpler elements until they reach more complex levels. Based on the studies of researcher Heraldo M. Vianna (1982) and the guidelines of Bloom *et al*.(1973), decisions are guided by objectives related to the corresponding evaluations.

Based on the assumption that "the lesson is the fieldwork of the teacher who investigates" and as support for drawing up the plan, Professor Dr Marisa Ramos Barbieri proposed Transversal Harmony (HT), one of the exercises used in her courses and workshops on teaching methodology and practice. At the end of the 1980s, HT began to be worked on with students from the Figueiredo Ferraz Institute (IFF-Fiocruz/RJ), who were taking part in "Saúde da Criança" (Child Health), a postgraduate programme in which "Research Methodology" and "Teaching Methodology" coexisted and were articulated. When the postgraduates were assessed, they realised the close relationship between research, guidance and teaching.

Considering the firm intention of imprinting an investigative character on teaching action, one of the procedures is to propose alternative plans, the elements of which are articulated by the teacher through horizontal harmony.

Here, this hierarchy of complexity, abstraction and difficulty is shown in the Tables below, known as Transversal Harmony, respecting the relationship between objective and assessment, in order to guide the teacher's thinking and anticipate the student's thinking.

Chart 1 - Initial planning stage for building general concepts.

Theme	**Specific content**	**Objectives\criteria**	**Reviews**
Ecology	Life cycle	Identify the stages of	Student

		holometabolous development	demonstrations;
Zoology	Arthropods	Recognise morphological adaptations	
Ecology	Ecological interactions	Recognise the trophic guilds of pollinators and the insect-plant relationship	

Chart 2- Stage of construction of specific and complex concepts.

Theme	Specific content	Objectives\criteria	Reviews
Evolutionary ecology	Holometabolous development	Relate holometabolous development;	Student demonstrations
	Parental care	Understanding that parental care guarantees differentiated survival parents choose food and a safe place	
	Eating behaviour	Investigate the history of holometaboles and hemi-metaboles and conclude that in holometaboles there are selective pressures to obtain food and protect offspring	

Chart 3 - Stage in the construction of a new, complex investigative perspective on the subject.

Theme	Specific content	Objectives\ criteria	Reviews
Biological	• Darwinian function:	Evolutionary	Publication

Adaptation with a Variational Perspective	assumption that all or many organic structures are involved in the Survivale differentiated reproduction of organisms, and respond primarily to the demands of the ecological environment. • Population thinking.	perspective on explaining adaptation.	in 2018 of an article in the journal Ciência em Tela.

Bibliography:

Aguiar, C. M. L. & Garófalo, C. A. 2004. Nesting biology of *Centris (Hemisiella) tarsata* Smith (Hymenoptera, Apidae, Centridini). Revista Brasileira de Zoologia, 21(2):477-486.

Alves dos Santos, I. Knowledge and breeding of solitary bees, a challenge. *Revista Tecnologia e Ambiente,* v.10, n.2, p.99-113. 2004.

Anderson, W. R. 1979. Floral conservatism in Neotropical Malpighiaceae. Biotropica, 11:219-223.

Andrietti, F.; Martinoli, A. & Rigato, F. 1997. Quantitative data concerning the oviposition of *Bombyliius fimbriatus* Meigen (Diptera:Bombyliidae), a parasite of *Andrena agilissima (*scopoli) (Hymenoptera:Andrenidae). Entomologists Record, 109:59-63.

Ausubel, D. P. Meaningful learning: David Ausubel's theory. São Paulo: Moraes, 1982.

Barbieri, M. R. Teaching Science in Schools: an open question. *Em aberto,* Brasília, n. 40, pp. 17-24. 1988.

Barbieri, MR The construction of teacher knowledge. Ribeirão Preto: Holos, 2001.

Bloom, B.; Krathwohl, D. R.; David, R.; Masia, B. B. Taxonomy of educational objectives: cognitive domain. Porto Alegre: Globo, 1973.

Bohart, R. M. & Menke, A. S. 1976. Sphecid Wasps of the World. A Generic Revision. Berkeley, Los Angeles, London: University of California Press. 1 colour plate, IX + 695 p.

Brodeur, J. & McNeil, J. N. 1989. Seasonal microhabitat selection by an endoparasitoid through adaptive modification of host behaviour. Science, 244:226-228.

Brodeur, J. & McNeil, J. N. 1990. Overwintering microhabitat selection by an endoparasitoid (Hymenoptera: Aphidiidae): induced phototactic and thigmokinetic responses in dying hosts. Journal of Insect Behaviour, 3:751-63.

Brodeur, J. & McNeil, J. N. 1991. The effect of host plant architecture on the distribution and survival of *Aphidius nigripes* (Hymenoptera: Aphidiidae). Redia, 74:251-58.

Brodeur, J. & McNeil, J. N. 1992. Host behaviour modification by the endoparasitoid *Aphidius nigripes-.* a strategy to reduce hyperparasitism. Ecological Entomology, 17:97-144.

Brodeur, J. & Vet, L. E. M. 1994. Usurpation of host behaviour by a parasitic wasp. Animal Behaviour, 48:187-192.

Brodeur, J., Boivin, G., 2004. Functional ecology of immature parasitoids. Annual Review of Entomology, 49:27-49.

Buchmann, S. L. 1987. The ecology of oil flowers and their bees. Annual Review of Ecology, Evolution and Systematics, 18:343-369.

Burian, R. The epistemology of development, evolution, and genetics. Cambridge University Press. 2005.

Cane, J. H.; Eickwort, G. C.; Wesley, F. R.; Spielholz, J. 1983. Foraging, grooming, and mating behaviours of *Macrospis nuda* (Hymenoptera: Melittidae) and use of *Lysimachia ciliate* (Primulaceae) oils in larval provisions. American Midland Naturalist, 110:257-64.

Camillo, E. 2005. Nesting biology of four *Tetrapedia* species in trap-nests (Hymenoptera:Apidae: Tetrapediini). Revista de Biologia Tropical, 53: 175-186.

Carvalho, A. M. C. 1990. Study of the interactions between apifauna and bee flora in cerrado vegetation -Panga Ecological Reserve - Uberlândia -MG. Master's dissertation, University of São Paulo.

Carver, F. J.; Gilman, J. L. & Hurd, H. 1999. Spermatophore production and spermatheca content in *Tenebrio molitor* infected with *Hymenolepis diminuta.* Journal of Insect Physiology, 45:565-569.

Combes, C. 2001. Parasitism. The Ecology and Evolution of Intimate Interactions. The University of Chicago Press, 728 pp.

Couto, R.M. & E. Camillo, 2014. Oil deposits by females of *Centris analis* (Fabricius) (Hymenoptera: Apidae: Centridini) parasitised by *Physocephala* spp. (Diptera: Conopidae). EntomoBrasilis, 7 (2): 8185.

Couto R.M; Castro, D.R.; Trigo, F.R; Perticarrari, A.; Barbieri, M.R. Teacher and journalist at school: production and dissemination of knowledge. Raízes e Rumos. v. 2, n. 2, p. 75-84, 2014.

Coville, R. E. & Coville, P. L. 1980. Nesting biology and male behaviour of *Trypoxylon {Tiypargllum) tenoctitlan* in Costa Rica (Hymenoptera: Sphecidae). Annals of the Entomological Society of America, 73:110-119.

Coville, R. E. 1982. Wasps of the Genus *Trypoxylon Subgenus Trypargilum in* North America (Hymenoptera: Sphecidae). University of California Press, Berkeley.

Coville, R. E. & Griswold, C. 1983. Nesting biology of *Trypoxylon xanthandrum* in Costa Rica with observations on its spider prey (Hymenoptera: Sphecidae: Araneae: Senoculidae). Journal of the Kansas Entomological Society, 56:205-216.

Coville, R. E. & Griswold, C. 1984. Biology of *Trypoxylon (Trypargilum) superbum* (Hymenoptera: Sphecidae), a spiderhunting wasp with extended guarding of the brood by males. Journal of the Kansas Entomological Society, 57:365-376.

Curio, E. 1976. The ethology of predation. Springer, Berlin Heidelberg New York.

Danks, J.V. Biology of some stem-nesting aculeate Hymenoptera. *Transactions of the Royal Entomological,* v. 122, p. 323-399. 1971.

De Jong-Brink, M.; Hoek, R. M.; lageweg, W. & Smit, A. B. 1997. Schistosome parasites induce physiological changes in their snail hosts by interfering with two regulatory systems, the internal defence system and the neuroendocrine system. In *Parasites and Pathogens: Effects on Host Hormones and Behavior* (ed. Beckage, N. E.), pp. 57±75. New York: Chapman & Hall.

Edmunds, M. 1974. Defence in animals. A survey of anti-predator defences. Longman, New York.

Endler, J. A. 1986. Natural selection in the wild. Princeton University Press, Princeton, N.J.

Evans, H. E. 1966. The behaviour patterns of solitary wasps. Annual Review of Entomology, 11:123.154.

Field, J. 1989. Intraspecific parasitism and nesting success in the solitary wasp *Ammophila sabulosa. Behaviour* 110:23-46.

Field, J .1992. Intraspecific parasitism as an alternative reproductive tactic in nest-building wasps and bees. Biology Reviews, 67:79-126.

Gazola, A. N. & Garófalo, C. A. 2003. Parasitic behaviour of *Leucospis cayennensis* Westwood (Hymenoptera, Leucospidae) and rates of parasitism in populations of *Centris (Heterocentris) analis* (Fabricius 1804) (Hymenoptera: Apidae: Centridini). Journal of the Kansas Entomological Society, 76:131-142.

Heinrich, B. & Heinrich, M. J. E. 1983. Size and caste in temperature regulation by bumblebees. Physiological Zoology, 56:553-567.

Hiller, B. & Wittmann, D. 1994. Seasonality, nesting biology and mating behaviour of the oil-collecting bee *Epicharis dejeanii* (Anthophoridae, Centridini). Biosciences, 2(1): 107-124.

Horton, D. R. & Moore, J. 1993. Behavioural Effects of Parasites and Pathogens in Insect Hosts. In: Parasites and Pathogens of Insects (Ed. by N.E. Beckage, S.N. Thompson, and B.A. Federici), pp. 107124. San Diego: Academic Press.

Hurd, P. D. 1978. An annotated catalogue of the carpenter bees (genus *Xylocopa* Latreille) of the Western Hemisphere. Smithsonian Institution Press, Washington, D.C. 106 p.

Hurd, H. & Arme, C. 1986. *Hymenolepis diminuta-,* influence of metacestodes on synthesis and secretion of fat body protein and its ovarian sequestration in the intermediate host, *Tenebrio molitor.* Parasitology, 93:111-120.

Kapil, R. P. & Dhaliwal, J. S. 1968. Defence of nest by the female of *Xylocopa fenestrata* Fab. (Xylocopinae, Hymenoptera). Insects Sociaux 15(4):419-422.

Katayama, E. & Maeta, Y. 1998. The fourth host of *Physocephala obscura* Krõber found in Japan (Diptera, Conopidae). Chugoku Kontyu, 12:23.

Klein, S. L. 2003. Parasite manipulation of the proximate mechanisms that mediate social behaviour in vertebrates. Physiology & Behavior, 79: 441-449.

Krasilchik, M. Prática de Ensino de Biologia. São Paulo: University of São Paulo Press, 197p. 2005.

Krombein, K. V. 1967. Trap-nesting wasps and bees. Life histories, nests and associates. Smithsonian Press; Washington, D. C.; vi + 570pp.

Linsley, E. G. 1958. The ecology of solitary bees. Hilgardia, 27:543-599.

Jayasingh, D. B. & Freeman, B. E. Trap-nesting solitary aculeates (Insecta: Hymenoptera) in St Catherine Jamaica. *Caribbean Journal of Science,* vol. 15, no. 3-4, p. 69-78. 1980.

Jesus, B. M. V. 1995. Nesting Behaviour and Seasonal Abundance of *Centris (Heterocentris) analis.* Master's dissertation - Ribeirão Preto School of Philosophy, Sciences and Letters - USP. 69pp.

Jesus, B. M. V. & Garófalo, C. A. 2000. Nesting behaviour of *Centris analis* in southeastern Brazil (Hymenoptera, Apidae, Centridini). Apidologie, 31(4):503-515.

Major, M.; Webb, T. J. & Hurd, H. 1997. Haemolymph from female beetles infected with *Hymenolepis diminuta* metacestodes retards the development of ovarian follicles in recipient *Tenebrio molitor* (Coleoptera). Parasitology, 114:175-179.

MacFarlane, R. P.; Lipa, J. J. & Liu, H. J. 1995. Bumble bee pathogens and internal enemies. Bee World, 76:130-148.

Mayr, E. Biologia, ciência única. São Paulo: Companhia das Letras, 2005.

Medler, J. T. 1958. Principies and methods for the utilisation of bumblebees in cross-pollination of crops In 10th Internatl. Cong. Ent. Proc., Montreal, Aug. 17-23, 973-981.

Michener, C. D. 2000. The bees of the world, The Johns Hopkins University Press, Baltimore.

Menezes, A. O. Jr. 1992. Taxonomic study of the Brazilian species of

Leucospidae (Hymenoptera, Chalcidoidea). Master's thesis. Federal University of Paraná. 177p.

Moore, J. 2002. Parasites and the behaviour of animals. In: Oxford Series in Ecology and Evolution. Oxford University Press, USA.

Minayo, M. C. O desafio do conhecimento: pesquisa qualitativa em saúde. Rio de Janeiro, HucitecAbrasco, 1992.

Muller, C.B. & Schmid-Hempel, R. 1992. To die for host or parasite? Animal Behaviour, 44:177-179.

Muller, C. B. & Schmid-Hempel, P. 1993. Exploitation of cold temperature as defence against parasitoids in bumblebees. Nature, 363:65-67.

Muller, C. B. 1994. Parasite-induced digging behaviour in bumble bee workers. Animal Behaviour, 48:961-966.

Múller, C. B.; Blackburn, T. M. & Schmid-Hempel, P. 1996. Field evidence that host selection by conopid parasitoids is related to host body size. Insectes Sociaux, 43:227-233.

Neff, J. L. & Simpsom, B. B. 1981. Oil-collecting structures in the Anthophoridae (Hymenoptera): morphology, function, and use in systematics. Journal of the Kansas Entomological Society, 54(1):95-123.

Otterstatter, M. C.; Whidden, T. L. & Owen, R. E. 2002. Contrasting frequencies of parasitism and host mortality among phorid and conopid parasitoids of bumble-bees. Ecological Entomology, 27:229-237.

Paxton, R. J.; Tengõ, J. & Hedstrõm, L. 1996. Dipteran parasites and other associates of a communal bee, *Andrena scotica* (Hymenoptera: Apoidea), on Õ land, SE Sweden. Entomologisk Tidskrift, 117:165- 178.

Pedro, S. R. M. 1994. Interactions between bees and flowers in a savannah area in NE São Paulo state: oil-collecting bees (Hymenoptera: Apoidea: Apidae). *In:* Meeting on Bees, 1. *Proceedings,* pp. 243-255.

Pereira, M.; Garófalo, C. A.; Camillo, E.; Serrano, J. C. 1999. Nesting biology of *Centris (Hemisiella) vittata* Lepeletier in southeastern Brazil (Hymenoptera, Apidae, Centridini). Apidologie, 30:327-338.

Polidori, C.; Disney, R. H. L. & Andrietti, F. 2001. Some behavioural observations on *Megaselia oxybelorum* Schmitz (Diptera: Phoridae), a newklepto parasiteof *Cerceris arenaria (*L.) (Hymenoptera:Sphecoidea:Philanthidae). British Journal of Entomology and Natural History, 14:93-95.

Poulin, R. 1994. The evolution of parasite manipulation of host behaviour: a theoretical analysis. Parasitology, 109:109-118.

Rasmussen, C. & Cameron, S. A. 2004. Conopid fly (Diptera: Conopidae) attacking large orchid bees (Hymenoptera: Apidae: *Eulaema).* Journal of the Kansas Entomological Society, 77:61-62.

Reitz, S. R. & Nettles, W. C. Jr. 1994. Regulation of *Helicoverpa zea* larval behaviour by the parasitoid *Eucelatoria bryani.* Entomol. Exp. Appl, 71: 33-39.

Roubik, D. W. 1989. Ecology and natural history of tropical bees. Cambridge, University, 514 pp.

Rizzini, C. T. 1997. Tratado de fitogeografia do Brasil. 2ª edition. Rio de Janeiro, Âmbito Cultural Edições Ltda, 747 p.

Sakagami, S. F. & Laroca, S. 1971. Observations on the bionomics of some neotropical xylocopine bees with comparative and biofaunistic notes (Hymenoptera, Anthophoridae). Journal of the Faculty of Science, Hokkaido University, Series VI, Zoology, 18:57-127.

Santos, A.M., J.C. Serrano, R.M. Couto, L.S.G. Rocha, C.A. Mello-Patiu & C.A. Garófalo, 2008. Conopid Flies (Diptera: Conopidae) Parasitizing *Centris (Heterocentris) analis (*Fabricius) (Hymenoptera:Apidae:Centridini). Neotropical Entomology, 37: 606608.

Schmid-Hempel, P. & Schmid-Hempel, R. 1990. Endoparasitic larvae of conopid flies alter pollination behaviour of bumblebees. Naturwissenschaften, 27:50-52.

Schmid-Hempel, P. 1991. The ergonomics of worker behaviour in social hymenoptera: I. Some preliminaries. Advances in the study of behaviour, 20:87-134.

Schmid-Hempel, R. & Muller, C. B. 1991. Do parasitised bumblebees forage for their colony? Animal Behaviour, 41:910-912.

Schmid-Hempel, P. 2001. On the evolutionary ecology of host-parasite interactions: addressing the question with regard to bumble bees and their parasites. Naturwissenschaften, 88:147-158.

Serrano, J. C. & Garófalo, C. A. 1978. Use of artificial nests for the bionomic study of solitary bees and wasps. Ciência e Cultura, 30:597-598.

Seigler, D. S.; Simpson, B. B.; Martin, C.; Neff, J. L. 1978. Free 3-acetoxy fatty acids in floral glands of *Krameria* species. Phytochemistry, 17:995-996.

Sepulveda, C.; Mortimer, E. F.; El-Hani, C. N. Investigations in Science Teaching 18.2: 439-479. 2013.

Silveira, F. A. & Campos, M. J. O. 1995. The mellissofauna of Corumbataí (SP) and Paraopeba (MG) and an analysis of the biogeography of Brazilian cerrado bees (Hymenoptera, Apoidea). Revista Brasileira de Entomologia, 39(2):371-401.

Simpson, B. B; Neff, J. L.; Seigler, D. 1977. *Krameria,* free fatty acids and oil-collecting bees. Nature, 267(5607): 150-51.

Slansky Jr, F. 1978. Utilisation of energy and nitrogen by larvae of the imported cabbageworm, *Pieris rapae,* as affected by parasitism by *Apanteles glomeratus.* Environmental Entomology, 7:179-185.

Sober, E. The Nature of selection: evolutionary theory in philosophical focus. Chicago: The University of Chicago Press. 1993.

Stamp, N. E. 1981. Behaviour of parasitised aposematic caterpillars:

advantageous to the parasitoid or the host? American Naturalist, 118: 715-725.

Tepedino, V. J.; Mcdonald, L. L.; Rothwell, R. 1979. Defence against parasitisation in mud-nesting Hymenoptera: can empty cells increase net reproductive output? Behavioural Ecology and Sociobiology, 6:99-104.

Thomas, F.; Schmidt-Rhaesa, A.; Martin, G.; Manu, C.; Durand, P.; Renaud, F. 2002. Do hairworms (Nematomorpha) manipulate the water seeking behaviour of their terrestrial hosts? Journal of Evolutionary Biology, 1: 356-361.

Thomas, F.; Adamo, S.; Moore, J. 2005. Parasitic manipulation: where are we and where should we go? Behavioural Processes, 68:185-199.

Webb, T. J. & Hurd, H. 1996. *Hymenolepis diminuta',* metacestode-induced reduction in the synthesis of the yolk protein, vitellogenin, in the fat body of *Tenebrio molitor.* Parasitology, 112:429-436.

Wcislo W. T. 1990. Parasitic and courtship behaviour of *Phalacrotophora halictorum* (Diptera: Phoridae) at a nesting site of *Lasioglossum figueresi* (Hymenoptera: Halictidae). Revista de Biologia Tropical, 38: 205-209.

Wcislo, W. T.; Wille, A. & Orozco, E. 1993. Nesting biology of tropical solitary and social sweat bees, *Lasioglossum (Dialictus) figueresi* Wcislo and L. (D.) *aeneiventri (*Friese) (Hymenoptera:Halictidae). Insectes Sociaux, 40:21-40.

Vianna, H. M. Test Planning. In: VIANNA, H. M. Testes em Educação. 4th ed. São Paulo: IBRASA, p. 29-47, 1982.

Viella, M. A. L. From intentions to educational objectives. In: CASTANHO, S.jCASTANHO, M. E. (Orgs.). Themes and texts in higher education methodology. Campinas: Papirus, p. 113-123, 2006.

Vinson, S. B. 1975. Biochemical coevolution between parasitoids and their hosts. In Evolutionary Strategies of Parasitic Insects and Mites (ed. Price, P. W.),pp. 14±48. Plenum Press, New York.

Vinson, S. B. & Iwantsch, G. F. 1980. Host regulation by insect parasitoids. Quarterly Review of Biology, 55:143-165.

V inson, S. B.; Frankie, G. W.; Williams, H. J. 1996. Chemical ecology of bees of the genus *Centris* (Hymenoptera: Apidae). Florida Entomologist, 79: 109-129.

V inson, S. B.; William, H. J.; Frankie, G. W.; Shrum, G. 1997. Floral lipid chemistry of *Byrsonima crasssifolia* (Malpighiaceae) and a use of floral lipids by *Centris* bees (Hymenoptera: Apidae) *Biotropica,* 29(1):76-83.

V inson, S. B. & Frankie, G. W. 2000. Nest Selection, Usurpation, and a Function for the Nest Entrance Plug of *Centris bicornuta* (Hymeoptera: Apidae). Annals of the Entomological Society of America, 93(2):254-260.

Vinson, S. B.; Frankie, G. W. & Williams, H. J. 2006. Nest liquid resources of several cavity nesting bees in the genus *Centris* and the Identification of preservative, levulinic acid. Journal of Chemical Ecology, 32:2013-2021.

Vogel, S. 1974. Õllbumen und Õlsammelnde bienen. Akad. Wissenschaften u. Literature Mainz. Franz Steiner, Wiesbaden, 547 PP-

Vogel, S. 1976a. *Lysimachia-.* Olblumen der Holarktis. Naturwissenschaften 63: 44-45.

Vogel, S. 1976b. Oil-collecting bees of the old World and their flowers. Abst. 15th International Entomol. Congress, Washington, DC. Aug. 1976.

Vogel, S. 1988. Die Õlblumen-Symbiosen: Parrallelismus und andere Aspekte ihrer Entwicklung in Raum and Zeit. Z. Zool. Syst. u Evolut. Forsch. 26: 341-362.

Vogel, S. 1990. The role of scent glands in pollination: On the structure and function of osmophores. (Translation:Smithsonian Instit. Libraries;

Amerind Pub., New Delhi). Rotterdam, Balkema, 202 pp.

MIX
Papier aus verantwortungsvollen Quellen
Paper from responsible sources
FSC® C105338

Printed by Books on Demand GmbH, Norderstedt / Germany